JETLINERS

Lance F Cole & John K Morton

Osprey Colour Series

JET LINERS
WINGS ACROSS THE WORLD

Published in 1990 by Osprey Publishing Limited
59 Grosvenor Street, London W1X 9DA

© Lance F Cole and John K Morton 1990

This book is copyrighted under the Berne Convention. All rights reserved. Apart from any fair dealing for the purpose of private study, research, criticism or review, as permitted under the Copyright Act, 1956, no part of this publication may be reproduced, stored in a retrieval system, or transmitted in any form or by any means, electronic, electrical, chemical, mechanical, optical, photocopying, recording or otherwise, without prior written permission. All enquiries should be addressed to the Publishers.

British Library Cataloguing in Publication Data

Cole, Lance F.
  Jet liners.
  1. Aeroplanes passenger commercial to 1985
  I. Title II. Series
  629.133'340423

ISBN 0 85045 955 9

Editor Tony Holmes
Designed by Paul Kime
Printed in Hong Kong

LANCE F COLE is an award-winning transport writer and aviation photographer. He specializes in capturing the art and emotion of flight and the firmament. He took his photographs for JET LINERS during a three year period spent flying the world's airways. Cole has flown over two-and-a-half million miles and spent many hours observing the magic of man's flight from the jump seat of a Boeing 747.

JOHN K MORTON has been photographing aircraft avidly for the past 20 years, his collection of civilian types featuring many classic moments in jetliner history. He has travelled across the globe capturing all manner of airlines operating aircraft in the most exotic of locations, his extensive photographic library totalling almost 10 000 transparencies. A transport buff from way back, John cut his 'photographic teeth' on the steam locomotives running on British railway lines in the 1950s and 60s, before turning to aviation.

All the photographs featured in JET LINERS taken by Lance Cole were shot with Olympus and Canon cameras loaded with Ilfochrome, Kodak KR and Fuji films.

John K Morton used a Pentax Spotmatic 1000 loaded with Kodachrome 25 film to photograph the earlier aircraft featured in JET LINERS. The more recent material was again shot with Kodachrome 25 film, this time loaded into a Minolta 9000 camera.

Lance would like to thank the staff at Koninklijke Luchtvaart Maatschappij KLM Royal Dutch Airlines for being the very, very best, and the authors express their gratitude to Japan Airlines, Qantas, British Airways and Lufthansa. Acknowledgement must also go to the various airport authorities around the world who provide facilities for the enthusiast and photographer to pursue their hobby, without which many of these photographs could not have been taken.

Front cover A gloriously colourful Aero Mexico D-9 Series 32 gleams as it approaches Miami International Airport in December 1987. This colour scheme has only appeared on Aero Mexico aircraft in the past two years

Back cover Wearing the famous Flying Tigers T on its tail, a highly polished natural metal Boeing 747F taxies out at Kingsford Smith Airport. The unmistakable Sydney skyline forms an impressive backdrop for the departing cargo Jumbo

Title Pages Cathay's backyard. The heat haze from the 747-300 auxiliary power unit shoots back over a trio of TriStars. Every aircraft in the picture is Rolls-Royce powered

Right A very tatty United 727-100 rests between uplifts at Chicago. This aircraft has obviously seen it all! Note the stretched DC-8 still in service alongside the newer 767 that can be seen tracking out to the runway. The 727 has sold in its thousands, it being the workhorse of the domestic American market and, as you can see, these machines are made to earn their keep. Ground time is kept to a minimum and six, or even seven flights a day are not unusual

Introduction

Within the pages of JET LINERS you will find some of the finest and rarest civil aviation images to be collected under one title. JET LINERS takes you on several different journeys. It features many rare shots of the jet age in both its first and second generations. The story also takes you into the heart of modern airline operations the world over.

The lives of the 707, the DC-8, the Caravelle, the VC10, the colours and travels of the Convair jets, and many more early civil jets are chronicled through the classic images of John K Morton. The life and times of airliners around the world can be seen through the photos of the much travelled Lance F Cole. The mighty 747s are shown earning their keep, the flying leviathans that circle the globe, the DC-10, Tristar, and best selling Airbus, are all portrayed.

The action covers the domestic American scene as well as Europe, the Far East and the Soviet Bloc. The world's great airlines, as well as those less well known, are featured as they have been seen around the world. Open the pages of JET LINERS and feast upon a sky full of great air transport moments.

Right Having offloaded their sun-seeking passengers a Balair DC-9, SAT Caravelle and a Hapag Lloyd 727 receive the attention of the cleaning staff at Corfu in August 1983. All three airlines deal mainly with charter work, Balair being based in Switzerland, and SAT and Hapag Lloyd in West Germany

Contents

Boeing classics 8

Passengers' perspective 20

Wide-bodied 24

British craftsmanship 68

The workhorses 82

New generation 102

European flavour 116

Limited editions 122

Boeing classics

Left Quito, Equador, is the base of Ecuatoriana, an airline that must carry one of the most attractive and colourful liveries in South America. This 707, HC-BFC, is on final approach to Miami in December 1987

Below The setting sun adds a sparkle to Transavia 707 PH-TVA parked at Manchester in October 1973. This Dutch charter airline now operates a fleet of Boeing 737s

Left Kenya Airways was formed in 1977 after the collapse of East African Airways. The newly formed carrier was using 707 5Y-BBK for its Nairobi to London service, here diverted to Manchester in December 1977 owing to fog at Heathrow Airport

Below Boeing 720 OY-APZ was one of several 720s operated by Danish charter airline Maersk Air. Pictured at Las Palmas in May 1976, the airline no longer flies this type

Overleaf Prior to the delivery of Boeing 747SP and 767 aircraft, Air Mauritius utilized 707s on their services to the United Kingdom. This November 1984 shot shows 3B-NAF at Terminal 3, London Heathrow

Above TAP Air Portugal 707 CS-TBD was employed on a holiday charter service out of Manchester International Airport in June 1983

Right An American Trans Air 707 rests at Las Vegas Airport in August 1981 after disembarking its passengers, whose destination is no doubt the gambling casinos

Above An unmarked EL AL 707-320 operates from Amsterdam amid very tight anti-terrorist security. As old 707s go, this one looks very well cared for

Left British Midland at one time had several 707s in its fleet but alas no more. In July 1974, G-AYVE is being prepared at Manchester for a flight over the North Atlantic

Left American Airlines no longer fly any 707 variants. N7568AA has just arrived at New York's John F Kennedy Airport and is taxiing to the freight terminal in August 1979

Top Nairobi Airport in August 1975 is the setting for Boeing 720 ET-AFB of Ethiopian

Above Air Malta Boeing 720 AP-AMJ is taxiing for take off from Manchester in June 1975. The aircraft was on lease from Pakistan International Airlines at the time and still wore its Pakistani registration as a result

Passengers' perspective

Left The North Pole from the air. Winging down to Paris from Anchorage, the cross roads of the air

Below Dawn over Siberia. The JAL Flight 402 glows over the ice. Those Pratt & Whitneys just kept on churning through the non-stop flight to Tokyo

Above On board Singapore Airlines ranging up the map, across the South China Sea and on into Hong Kong. The highly polished cowlings were almost blinding in their intensity

Right Dawn filters in over the northern Pacific. Lance Cole flew into Anchorage on the jump seat of this KLM 747-300. In company with Captain P K Lodder, First Officer B Wowbers who flew the aircraft, Engineer P Defaix and Purser L Baggen, he watched the dawn break and daylight engulf the Alaskan mountains. In a display of sheer professionalism the flight from Tokyo to Anchorage took exactly the five hours 59 minutes predicted by the inertial navigation system. Many thanks to the flying Dutch for a truly great flight on a great airline

Above High over central Siberia the non-stop Japan Airlines Flight 402 from London to Tokyo takes just over 11 hours and sweeps across the map in style. The frozen wastes seem to go on for ever

Wide-bodied

Left Hong Kong's spectacular skyline shows off this Cathay Pacific 747-300 to dramatic effect. The people living in the high rise blocks are probably not aviation enthusiasts!

Below Jumbo Art. British Airways 747 seen alongside a South African 747-300 Extended Upper Deck machine. The sharper, taller tail in the background belongs to the short bodied 747SP operated by the United Arab Emirates Government

25

At 16 degrees Centigrade below freezing, Anchorage is positively warm as two Korean Air 747-200s are separated by a JAL machine. The frozen ramp may account for the lack of workers

Above A Flying Tigers 747F slithers over the ice at Anchorage, the mountains which border the airport forming an impressive backdrop to the silver freighter. Notice the insect-like prop craft huddling behind the big Boeing

Right Cruising in over Kowloon, the prestigious British Airways Flight BA 20 approaches Kai Tak. Under the setting oriental sun the empire strikes back with the roar of four Rolls-Royce RB.211-524D4s and the fleeting flash of blue and silver

Above Sunset Boulevard? Two Japan Airlines 747s get towed into the setting oriental sun. These two are just part of the biggest 747 fleet in the world. JAL currently operates 59 747s. They are as follows: eight 747-300s, 23 747-200s, nine 747-200Fs, 13 747SR high density specials, five 747-100s and one 747-100F. They have 20 747-400s on order and one of the 747 SRs has had the galleys removed for local domestic routes, the aircraft now seating 563 passengers

Right Cathay Pacific's vivid colours dominate a very busy Hong Kong Kai Tak whilst a United 747SP and a Swissair long range DC-10 bake under the searing sun

Above It is below freezing as two JAL 747s shiver on the ramp. At the rear is an early 747-100 and in front a -200, the difference can be seen here in the upper deck windows. Also visible is Narita's new runway. It is so long that reverse thrust is nothing more than a fleeting hiss. The angled high speed taxiways cut down on undercarriage wear and also permit staggered take-offs

Right JAL's newest 747-300 blazes under the blinding lights while a -200 series parked behind lights up prior to departure. Photographed from high above, these giants appear toy-like in scale

The classic shot of Hong Kong

Above and right This view of the 747-300 really does capture the giant size of this leviathan. A ground worker can be seen under the wing, showing the scale of Mr Boeing's beast. The KLM colours seem to elongate the fuselage. Visible in the background are the double airbridges—a first for Amsterdam Airport

Overleaf Architecture and the aeroplane. A United 747-100 lurches up, up and away from Narita whilst the 'night sun' ramp illuminations look on

Left Resplendent in its newly acquired livery, British Airways 747 G-AWNO City of Durham is pushed back from the gate at Manchester in February 1985

Above Closely followed by one of the carrier's L-1011 TriStars, TWA Boeing 747SP N57202 is turning on to runway 24L at Los Angeles International in August 1981

Above Barely six months old when photographed on the approach to Los Angeles in August 1981, Qantas 747SP VH-EAA is one of two special performance Jumbos operated by the airline

Right The sweeping lines of the Avianca livery appear to suit the 747 rather more than any other type of aircraft. This August 1979 shot of HK-2000 was taken at Miami as the Jumbo was taxiing to the end of the runway for take-off

Right Although wearing Metro International titles on its tail, this 747 actually belongs to Flying Tigers. The aircraft is seen here at Los Angeles International Airport in August 1981 receiving Metro transfers to its fin. Metro International is in fact a Flying Tigers subsidiary

Left Virgin Boeing G-VIRG spent part of its life flying in and out of Buenos Aires when in the fleet of Aerolineas Argentinas. Today its destinations are New York and Miami. Gatwick Handling is preparing the 747 for departure in August 1985

Below Two giants rest between work. A Canadian Airlines DC-10 holds station beside one of the Flying Tigers fleet. The similarity between the new Canadian livery and that of British Airways is obvious in this shot. Maybe Canadian should have stayed with their famous orange image?

Left 'Fly the silver fleet', American Airlines' aircraft have been without paint for decades. This DC-10 is departing Frankfurt for Chicago's O'Hare field. Note the tail fin spars which are clearly defined. In the foreground a Pan Am 747 awaits pushback, while behind lies a BWIA TriStar

Above The new colours of the Malaysian Airlines System, now known rather curiously as 'Malaysia'. This DC-10 received a wash and brush up while on the ramp at Tokyo Narita

Left Blast off!! A Bangledesh Biman DC-10 soars away from Heathrow's Runway 27 Right on a winter evening. This shot clearly shows the over-sized tail plane, which is actually smaller on the new MD-11 derivative, and the huge main gear doors

Above Lufthansa at home. A DC-10 awaits taxi clearance at the Flughafen Frankfurt. Unlike many airport ramps Frankfurt's is continually scrubbed-clean. The typical Douglas angled nose wheel is clearly shown. Even the DC-6 had this feature!

Northwest Orient is in the process of deleting the word 'Orient' from its titles, and are now known purely as 'Northwest'. In its original form, DC-10 N151US is landing at Miami in October 1984

Above Balair is a subsidiary of Swissair and provides charter flights to various destinations. DC-10 HB-IHK is arriving at Los Angeles West Imperial charter terminal in August 1981

Left The aggressive lines of the Douglas DC-10 are graphically illustrated in this study. The aircraft is one of five convertible DC-10 CFs that Sabena operate. The cargo cabin is located ahead of the wings

Above Formerly known as British Caledonian Charter, this Cal Air DC-10, G-BJZD, is seen in the final livery used by the airline. The aircraft is fully kitted out as an economy class only machine, and is seen arriving at Manchester in August 1986

Right Also configured as an all-economy class aircraft, Cal Air DC-10 G-GCAL is seen just before touchdown at Ibiza in August 1987

Above Rio de Janerio is the location for this April 1983 shot of Lan Chile's DC-10 CC-CJT. Lan Chile no longer fly this type and the aircraft now carries the livery of American Airlines

Opposite above The polished fuselage of Aero Mexico DC-10 XA-DUH catches the evening sun as it turns on the runway at Miami in April 1980

Opposite below An Eastern Airlines 727 follows Viasa DC-10 YV-135C on the taxi-way at Miami in January 1988

Left The beast rolls in. For many, the DC-10 is the most aggressively photogenic widebody jet. Here an American Airlines DC-10 howls its way across the ramp at Toronto on a freezing winter's evening. Hiding behind that stylish angled tail is a Wardair A310

Above The place is Anchorage, capital of Alaska. The time is 0530 hours. The sunlight is streaming through and warming the mountains in the background. Meanwhile a very grubby SAS long range DC-10 Series 30 is fed and watered. SAS were pioneers of the polar route using Douglas equipment. This example of their modern all DC-10 long haul fleet has only been half washed, hence the grime behind the wing; it must have been some storm!!

This KLM Royal Dutch Airlines DC-10 has just arrived from Lagos via Kano—as you can see from the paintwork—or lack of it! Named *Maurice Ravel*, and registered PH-DTD, it obviously needs a respray

Above One of the 17 JAL DC-10 Series-40s is attended to at Narita whilst an Iraqi tail pokes out from the cargo bay. JAL use their DC-10s on medium length routes and on southern Pacific sectors

Right A water reflective shot taken when a Malaysian DC-10 was scrubbed clean on the ramp at Narita

Above A tail of two cities? An Air Lanka TriStar fresh in from Colombo meets up with a sister airframe in the form of an All Nippon Airways example at Narita

Right Lockheed L-1011 TriStar C-GIFE operated by Canadian carrier Worldways arrives at Manchester in June 1988 on one of its many flights between Toronto and the UK

Hawaiian Air entered into the wide-bodied league during 1985 with five L-1011 TriStars originally flown by All Nippon Airlines of Japan. N763BE is landing at Las Vegas in August 1988 at the end of a scheduled flight from Honolulu

The workhorses

Below Connected to the American Airlines' passenger terminal by a boarding ramp, an AA MD-82 is prepared before receiving passengers at Chicago O'Hare International Airport. American Airlines operates a staggering 150 MD-82 aircraft and has placed an order for a further 70 from McDonnell Douglas. This particular MD-82 is bound for Nashville, Tennessee

Right Manchester International Airport is again being used to handle diversions from a fog bound London Heathrow. DC-9 OE-LDN is approaching the gate following its flight from Austria in December 1977

Below Midway Airlines DC-9 N1062T is waiting to receive instructions that will see it pushed back from its stand at La Guardia airport, New York, in June 1981

Right American Airlines' own ramp at Chicago is a very crowded place. Here an MD-83 arcs out in unison with a 727-200. The tail of an AA DC-10 Series 10 towers over them both

Left Bahamasair, the national airline of the Bahamas, currently has a total of four Boeing 737 aircraft in its fleet, and C6-BFC is pictured about to land at Miami in December 1988

Above Western Airlines, originally a major North American carrier based in Los Angeles, was merged with Delta Airlines during the early part of 1987. This 737 has just turned on to the runway at Las Vegas in April 1985

Left This Continental 737 carries the registration of its previous owner, People Express, and is arriving at its stand at Jacksonville airport, Florida in April 1988. People Express, servicing North America and Europe, merged with Continental Airlines in early 1987

Above Air Europe is a British airline with a fleet of Boeing 737s and 757s. This study of a 737 tail was taken at Manchester in September 1981

Above Although owned by Universair, 737-300 EC-EDM is seen at Ibiza in August 1987 wearing the colour scheme of Las Vegas based airline Sunworld. This Spanish charter airline, formed in 1987, is based on the island of Palma

Above German charter airline Condor flies to most Mediterranean destinations. This 737, D-ABHT, is about to take-off from the island of Ibiza in August 1987

Left In the background an original Heinkel 111, below one of Pan Am's European 727-200 fleet. An odd contrast of images found at Frankfurt

Above With the slats and flaps already pulled in, an Air Canada 727-200 streaks across the Toronto ramp. The thick wing chord and centre wing box are clearly shown in this head-on shot

Above The East River and Flushing Bay provide a pleasing background for this view of several Boeing 727s huddled together at the TWA domestic terminal at La Guardia airport, New York, in August 1979

Right Dawn at Bahrain revealed this Arania Afghan Airlines 727-100 parked with her three thrust reverse buckets hanging out. The soot-encrusted jet pipes indicate that the triple Pratt & Whitney JT8Ds are real oil burners!

Avianca, the Colombian carrier with its base in Bogota, has an all jet fleet of Boeing aircraft. Here, one of their 727s, American registered N203AV, is arriving at Miami in December 1987

Main picture As with all Federal Express aircraft this 727, N102FE, will spend most of the daylight hours on the ground, and depart to its destination overnight

Inset Caymen Airways has its base in Georgetown, Guyana, one of the myriad of islands found in the Caribbean. After the demise of American carrier Air Florida, one of the 727s originally flying in Air Florida colours was repainted in the colours of Caymen. The aircraft still carries the registration of its former owner, N272AF

Above Sterling Airways, a Danish charter airline, is seen using one of its 727s, OY-SBH, on a return leg flight from the island of Ibiza in August 1987

Right Kai Tak collage. Fresh from the paint shop, an Emirates 727-200 frames the busy ramp scene. Behind can be seen a Cathay TriStar, Thai A300 and a CAAC Boeing 767. Above them all can be seen the stunning colours of a Philippines 747-200 as it floats down to the 3331 metre Runway 13 Left

New generation

Left The stunning grey and gold of Alia, the Royal Jordanian Airline. This is their *Princess Iman*, an extended range Airbus A310-300 wingtip fenced model. The aerofoil section at the wing trailing edge can be seen dropping away from the starboard wing, as can the wing root slat fillet—typically advanced Airbus aerodynamics

Below Jordan's national airline, originally known as Alia, now carries the title Royal Jordanian on its aircraft. This study of the tail markings of Airbus A310-300 at Frankfurt in October 1987

Left A Turkish A310 speeds out towards the main runway at Frankfurt. The Thai 747-200 in the foreground is waiting to depart for Bangkok

Below Beautiful Bali. Amid the gardens of Bali's Denpassar airport, two Garuda A300s await the next load of tourists. These are early B4 series machines and await the new Garuda colour scheme

Overleaf Tokyo by night. A Philippines A300 gets ready to leave for Manilla. It's the last flight of the day and it's packed

Above The *Clipper Berlin* is captured departing from the Pan Am hub at Frankfurt. The full span leading edge slats can be seen powering out. The uninterrupted slat on a pylon equipped wing is a major Airbus achievement

Right On lease from Trans Australia Airlines for the European summer season, Airbus D-AITA is fully painted in the Condor livery and is seen leaving Ibiza with returning holidaymakers in August 1984

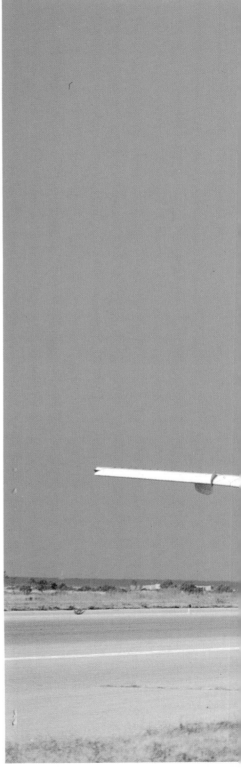

Above A Pratt & Whitney JT9D-7R4E1 powered A310 of Singapore Airlines trundles across the ramp at a very damp and windy Singapore Changi. SIA use their self-styled '3 TENS' on short and medium sectors around the spice islands. SIA operate a confusing amount of equipment to do the same job. Alongside the Airbus they have four Boeing 757-200s. Powerplants are standardized Pratt & Whitneys. This aircraft had just operated the combined SIA/Garuda service to Jakarta and Denpassar. The fuel saving from using the Airbus instead of a 727 or DC-9 is about 35 per cent

Above Hapag-Lloyd currently operate an all jet fleet of aircraft from their Hanover base. Airbus Industrie A300 D-AHLB has just received instructions from Corfu tower to depart for its German destination in August 1983

Overleaf Boeing 757 PH-AHF at Fort Lauderdale, Florida, in January 1989. The aircraft is in the colour scheme of Air Holland, its present owner. Odyssey International is a Canadian charter airline

Opposite above TWA is one of several national airlines regularly flying twin engined jets over the North Atlantic, although routes have had to be adjusted so that in case of emergency the aircraft is no more than 90 minutes flying time from a convenient airport. Here, Boeing 767ER (Extended Range) N610TW has just arrived at Frankfurt after completing a flight from St Louis via London Gatwick in October 1987

Opposite below Air Canada's extended range 767 swings up to the gate at Toronto's Lester B Pearson Airport

Above Northwest's *City of Portland* seen from above prior to pushback from Toronto. The sharp features of the elegant 757 are beautifully enhanced in this shot

European flavour

The early morning Air France flight to Paris is being prepared for departure from Manchester on a Saturday in June 1977. This routine service is being performed here by a Series 3 Caravelle

Above Resembling a scene from the late 1960s, this shot was in fact taken at Geneva in the late 1980s! Only recently have CTA put their superb Caravelle Series 10 aircraft out to grass

Opposite above With Jamaica Bay in the background, Polish Airlines Ilyushin Il-62M SP-LBB, having dispensed with the services of the push tractor, prepares to leave the Pan Am terminal at New York's John F Kennedy Airport in August 1980

Opposite below An Aeroflot Il-62 makes a rare visit to Frankfurt. The extra tail wheel support can be seen under the engines. Often billed as a VC10 copy, the machine is known to have some handling problems and requires a long, low approach pattern. Similarly, the take-off resembles horizontal linear levitation

Inset The Yugoslavian airline Aviogenex rely on a solitary Boeing 737 and a fleet of 727s and Tupolev Tu-134s to operate their charter services. Here Tu-134 YU-AJA is departing Manchester for its homeland in July 1975

Main picture Wearing Cubana colours a Tupolev Tu-154 departs Barbados in April 1984 on a scheduled flight to Havana

Limited editions

Inset In the early 1970s, charter journeys to Canada were made aboard Air Canada DC-8s. CF-TIS is just about to come to a halt at Manchester in June 1974 after completing a flight from Toronto

Main Picture Arrow Air has a fleet consisting entirely of DC-8 aircraft, this example, N441J, seen on approach to Miami in January 1989

Above Needless to say, 'Big A' is the call sign used by Arrow Air

Above Surinam Airways DC-8 Series 63 N4935C, built in 1968 is about to turn on to the runway at Miami in December 1988

Right Air Zaire 737 9Q-CNK heads the line up at Nairobi airport in August 1975. Also waiting to receive passengers for departing flights are East African Airways DC-9 5Y-MOI, Alitalia DC-8 I-DIWJ and Somalia 720 G-BCBB

Left Only a few more feet to touch down for Air Canada Cargo DC-8, CF-TJO, a Series 54 aircraft with Pratt & Whitney engines. Still operating cargo DC-8s into Heathrow today, this aircraft is seen approaching this airport in September 1976

Above Spantax, the Spanish charter carrier, had no less than twelve Convair Coronado 990 aircraft in its fleet at the time EC-BJC was photographed arriving at Las Palmas in April 1976 carrying a complement of holidaymakers

Overleaf As the sun sets in the west, an Eastern Airlines Lockheed L-1011 TriStar departs Miami in December 1987